The Muddy Dragon

by Sara E. Turner

There are two kinds of alligators. The American alligator lives in the United States.

United States
United States
China
Alligator regions
N
W
E
S

The Chinese alligator lives in China. The Chinese people call it the *muddy dragon*.

China
0 1,500 3,000 miles
0 1,500 3,000 kilometers

Where's the muddy dragon?

It's in its burrow.

A burrow is a kind of shelter. In winter, muddy dragons need shelter from the cold.

Where's the muddy dragon?

It's in the water.

It needs water to drink and keep cool. But it must come up for air. Muddy dragons need air to stay alive.

Where's the muddy dragon?

It's hunting in the pond.

It eats birds, snails, and small fish. Muddy dragons need food to stay alive.

Where's the muddy dragon?

Here it is!

And it's huge! It can grow to be 1½ meters (5 feet) long.

Muddy dragons have basic needs.
They need:

shelter
water
air
food